i

为了人与书的相遇

撰文　周宗伟　Text by Zhou Zongwei　　绘图　朱赢椿　Illustrations by Zhu Yingchun

蛛　嚅
SPIDER

广西师范大学出版社
·桂林·

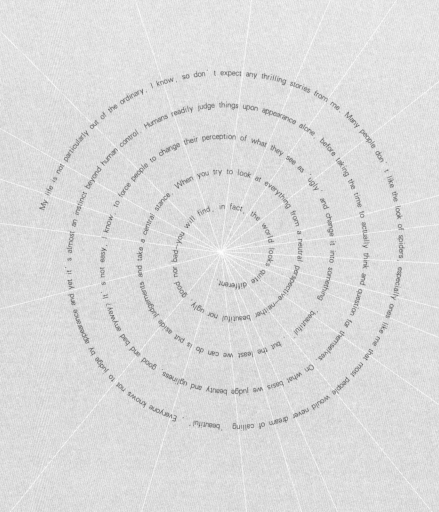

My life is not particularly out of the ordinary, I know, so don't expect any thrilling stories from me. Many people don't like the look of spiders, especially ones like me that most people would never dream of calling "beautiful." Everyone knows not to judge by appearances and yet it's almost an instinct beyond human control. Humans readily judge things upon appearance alone, before taking the time to actually think and question for themselves. On what basis we judge beauty and ugliness, good and bad anyway? It's not easy, I know, to force people to change their perception of what they see as "ugly," and change it into something "beautiful," but the least we can do is put aside judgements, good and take a central stance. When you try to look at everything from a neutral perspective—neither beautiful nor ugly, good nor bad—you will find, in fact, the world looks quite different.

我知道，我的生活普通又平凡。所以，别期待我会给你讲惊心动魄的故事。我也知道，很多人不喜欢蜘蛛的长相，尤其像我这样的蜘蛛去书房采集灰尘……但以貌取人几乎是人类难以控制的本能。人类总是不假思索地被各种事物的外表操控左右，却从没真正思考和探索过……但以貌取人，用相貌去判断事物，但要让人们把一个看起来"丑"的东西硬拔成"美"的，这很难做到。但是，人们至少可以先学习放弃这一切……不美，也不丑；不好，也不坏。你会……一切都看成中性的——不美，也不丑；不好，也不坏，以崭新的面目呈现。

那时候，我还是一只小蜘蛛。有一天，妈妈告诉我，每一只蜘蛛都要织自己的网，该怎么做，需要你独自寻找答案。于是，我离开了她，开始独立生活。我为了织网抛出的第一根蛛丝，在空气中摇摆，究竟要落向何方，还无法确定。

Back then when I was still a little spider, my mother told me one day that every spider needs to weave its own webs and that it is up to each spider to work out exactly how. So, when I left her I set off on my own journey. Casting out my first ever strand of silk to make a web, I watched as it swayed in the wind with no idea where it would land.

蛛丝飘来飘去，找不到它的方向。这时我开始有点想妈妈，真想她来帮我一把。

The strand of silk floated around in no clear direction. I began to miss my mother a little and wished she could help me.

·6·7· 蛛丝终于落到一截树枝上，总算有了依靠和着落，我大大松了一口气。

At last, the strand of silk landed on a branch offering some support at least. I let out a sigh of relief.

·8·9· 万万没想到，原来蛛丝缠住的根本不是枯树枝，而是一只伪装成树枝的尺蠖，他伸了一个懒腰，我的丝就断成两截。

It never occurred to me, that the branch my silk had attached to was actually no branch after all, but an inchworm in disguise. It stretched out its body and broke my silk in two.

·10·11· 我气呼呼地瞪了尺蠖 眼，只能从头再来。这次，我睁大了眼睛，终于找到真正的树枝，绑定了我在这个世界上的第一根蛛丝。

I stared angrily at the inchworm with no choice but to start all over again. This time, a little wiser, I found a real branch and I attached my first ever silk strand onto it.

· 12 · 13 · 我站在自己的蛛丝上，有些小小的得意。忽然一只蛞蝓爬了上来。
我不知道该高兴还是紧张，他究竟是猎物，还是敌人？

Standing on my own silk, I couldn't help but feel rather proud. Then suddenly, along came a slug.
I wasn't sure if I should be happy or afraid. Was it prey or predator?

· 14 · 15 · 蛞蝓离我越来越近，我到底该小心退后，还是大胆向前？但一根纤细的根蛛丝远远承受不了我和蛞蝓的重量，我们一起摔得很惨。

The slug edged towards me. Should I carefully retreat or bound forward bravely? There was no way the slender strand of silk could bear the weight of us both. We tumbled and fell into a sorry heap.

·16·17· 摔得很惨，让我学会把自己放低。我不再贪恋那些高高在上的树枝，这次选择从低处的细小草茎开始，不知不觉中很快就织好了一张网。

The nasty fall taught me to be more humble. I would no longer reach greedily for the high branches, instead this time I chose to explore the lower blades of grass and in no time at all I'd woven myself a web.

虽然织出的第一张网形状不太好看，但总算心血没有白费，似乎有飞虫自投罗网。可我兴奋地跑过去，发现原来只是几片花瓣。飞虫远比花瓣聪明，它们避开蛛丝，从网中央径直穿过。我空欢喜了一场，只好耐着性子再把蛛网补密。

My first web was a little pear-shaped, but it wasn't a complete disaster and some insects seemed to have flown into it already. Feeling excited, I rushed over to have a look only to discover merely a few flower petals. Insects are much smarter than petals, they know how to fly through the gaps and avoid getting stuck in the strings. I had got excited too soon, so I went about making the web more close-knit.

忽然刮起了一阵风，我的网开始随风飘荡，我不禁开心地在网中荡起了秋千。可风却越来越大，不仅刮断了草茎，也撕碎了我的网。我辛辛苦苦织成的第一张网还没粘到猎物，就这样随风而去。

All of a sudden there was a great gust of wind which swept up my web. I couldn't help but enjoy swinging along with it at first, but then the wind grew stronger and stronger, until eventually it broke the blades of grass my web was attached to and the web itself. I hadn't even caught anything in this web I had so painstakingly woven before it was blown out in the wind.

· 22 · 23 · 我吸取了教训,这一次在牢靠的树枝上织好了新的网。我还有了个邻居叫"悦目金蛛",他可以把网织成漂亮的字母图案,他的身上处处透着光芒。我心里有些自卑,一度想变成他。

Learning from this experience I made another web, this time attaching it to a sturdy branch. I had a neighbour there too, a priest spider, who made wonderful webs decorated with spectacular lettering and a body that seemed to glow all over. I felt rather inadequate and longed to be him.

忽见一只大鸟俯冲而来，悦目金蛛还没反应过来就已丧身鸟嘴，只留下破碎的字母蛛网在空中飘荡。我在旁边吓出一身冷汗，却也庆幸不起眼的外表让自己逃过一劫。

Suddenly a great big bird swooped in and before the priest spider could even react he was crushed to death in the bird's beak, leaving all but a web of letters floating in the air. I broke out into a cold sweat, although secretly I was glad my ordinary appearance had saved me.

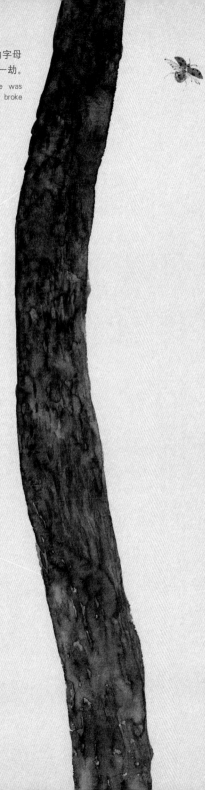

悦目金蛛的死还是让我心有余悸，我变得失魂落魄，毫无目的地到处游荡。我走累了，打算在一根枯竹竿上织网，暂时在这里安一个简陋的家。

The death of the beautiful priest spider left me with a lingering fear, I walked until I was completely exhausted, then decided to make a web on a stick of dead bamboo and set up camp there for the time being

我刚准备抽丝织网，竹管里突然窜出几只蜘蛛，对我发出凶狠的警告。原来我误入了漏斗蛛的地盘。他们会躲在洞穴中，织出像漏斗一样的网，等待不知情的猎物上门。我知道误闯了是非之地，赶紧逃离。

Just as I began to draw silk, a few spiders crawled out of the bamboo stalk, giving me a stark warning. It turned out I had stumbled into the territory of some funnel-web spiders. These spiders live in holes where they weave webs like funnels and await a visit from unsuspecting prey. I realised this was dangerous territory and made a quick escape.

我找到一只蘑菇，老老实实织了一张小网，眼巴巴地守网待虫。远处飞来一只苍蝇，看不起我这个普普通通的小蜘蛛，趴在高处嘲笑我那可怜的小网。我明明很饿，却奈何他不得。

I found a mushroom and made a small web beneath it, then waited anxiously for some insects. A fly flew in from the distance. He looked down on me scornfully, just an ordinary spider. He laughed at my pitiful web from his high up perch. I was starving, but the fly was no use.

上帝和苍蝇开了个玩笑，他在苍蝇背后悄悄安插了一只跳蛛。跳蛛俗名"蝇虎"，是捕猎苍蝇的能手，他们不用织网，可以靠迅猛的速度和力量捕杀猎物。苍蝇忘情地嘲笑着我，却没防备到背后跳蛛的突袭，在笑声中毙命。

This time, god played a trick on the fly by quietly placing a jumping spider behind the fly's back. Jumping spiders, also known as 'fly tigers', are expert fly catchers. They don't need to make webs, instead they use their immense speed and power to catch prey. The fly was too busy laughing at me that he didn't think to check his own back and just as he let out a cackle of laughter he met his sorry fate.

看到跳蛛如此凶猛，吓得我赶紧躲到蘑菇下面，大气也不敢出。原以为跳蛛有了食物就会离开，却没想到又凭空杀出一只蛛蜂，一针让跳蛛毙命。

Shaken by the ferocity of the jumping spider, I hid under the mushroom, breathing heavily and too afraid to go out. I assumed it would be safe now and that the spider would leave with its tummy full. I certainly did not expect that a spider wasp would appear out of thin air and kill the jumping spider with one fatal sting.

恐惧让我只想逃跑。我仓惶逃离了蘑菇地，误打误撞遇见了一片池塘。我看到对面竟然有一只蜘蛛正在用前肢有节奏地轻轻拍打水面，吸引小鱼游到他跟前。

All the fear made me want to escape. I fled from the mushroom and found myself at a pond. On the opposite side of the pond a spider was tapping rhythmically at the water with its front legs, drawing the small fish towards the surface.

我几乎揉痛了自己的眼睛，终于知道这不是在梦里。原来世上还有一种蜘蛛叫"鱼蛛"，是我们蜘蛛家族里的渔夫。他们和跳蛛一样也不织网，而喜欢在水边靠捕鱼为生。我不妄想和他一样去捉鱼，但越发纠结自己该往哪里去。

Rubbing my eyes in disbelief, I realised I was not dreaming. There is indeed a kind of spider known as the 'fish spider' —the fishmonger of our family. Like jumping spiders, they too do not make webs, instead they catch fish at the edge of the water to survive. I had no intention to catch fish in the same way, but it made me wonder where should I go.

长久地等待和守候是我们这类蜘蛛最常规的生活，但屡战屡败的经历让我变得躁动不安，我忽然产生了冒险的冲动，也许要活动起来才会有更多的机会。我玩了点聪明，把网织在了山羊角上，行走的山羊会给我带来更多机遇，而我却可以不费力气。

These spiders spend all day waiting and expecting, but in my case the constant cycle of battle and defeat had made me restless. I was suddenly compelled to take a risk. Perhaps by doing so it would open up more opportunities. In a rather clever move, I constructed a web between the horns of a goat so that when it walked, I had greater chances to catch things without the need for so much effort.

我躺在山羊角上，幻想某天一睁开眼，蛛网上就粘满了各种肥硕可口的猎物。风中偶尔飘来的小飞虫给我打了牙祭，但还没喂饱肚子，就发生了意外。山羊和同伴打架，我在山羊角的碰撞中左躲右闪，不仅蛛网被弄破，还险些被羊角刺伤。我一边责怪山羊暴躁，一边仓皇逃跑。

Resting on the goat's horn, I imagined a day when I would open my eyes and find a web filled with all sorts of delicious juicy prey. The occasional flying insect made for a welcome treat, but before I had eaten my fill, something unexpected happened. The goat broke into a fight with another animal of its kind. I dodged left and right avoiding blows between the two goats' horns. My web was destroyed and I only narrowly escaped being stabbed myself. I cursed the goat for its aggressive behaviour and ran away in a hurry.

我知道，除了从头开始，我别无选择。织网和捕食是我们这类蜘蛛生命中无法打破的循环。一次次地失败，一次次地从头再来，而等待的过程中又有无数的不确定。

Once again, I had no choice but to start from scratch. Making webs and catching prey is a never—ending cycle in the life of us spiders. One failure after another, starting over again and again and you never know what might happen as you wait.

一只蜗牛爬上了我辛辛苦苦拉好的第一根丝，立刻唤醒了我和蛞蝓的那段不愉快经历。我想赶走她，但已经来不及。只见她优雅从容地走在我的蛛丝上，宛如一名技巧高超的杂技演员，我惊叹她如何保持了平衡。

A snail began to edge along the first strand of silk I had so arduously put in place, reminding me at once of the slug fiasco. I wanted to shoo her away but it was already too late. I watched as she gracefully glided along the silk wire like an expert acrobat and marvelled at her unwavering balance.

这一次，我和蜗牛共同创造了一个奇迹——她成功地走完了蛛丝，并保持了蛛丝的完整。我的辛苦没有白费，也从蜗牛那里学会了平衡的智慧。

Together the snail and I had created something miraculous as she managed to walk the length of my silk, all the while keeping it completely intact. My efforts were not wasted and I learnt something about balance from the snail.

· 50 · 51 · 蜗牛告诉我，大部分飞虫都喜欢光，每到晚上，灯下就会聚集很多的飞虫，这里是个适合织网捕
猎的好地方。我决定在这里长久驻扎。

The snail told me that lots of flying insects are attracted to light and at nighttime they often gather around
lamps. It seemed like a good place to catch some prey, so I decided to stay there for a while.

蜗牛的建议果然有效，我尝到了甜头，在这里得到平生最大的收获。可没想到，知道这里飞虫多的并非只有我一个，早有壁虎在旁边觊觎。在网上挣扎的小飞虫，引来壁虎垂涎三尺。他不仅盯着我网上的飞虫，也虎视眈眈地盯着我，我只得忍痛弃网逃生。

The snail was absolutely right, I had a taste of success with the biggest catch of my life. It hadn't occurred to me however that I was not the only one and it turned out to be a coveted spot for geckos. The insects that struggled in my web attracted a gecko who came along salivating with envy. He stared at the insects in my web and glared at me fiercely, I had no choice but to abandon my web and make a run for it.

在灯下待久了，我已经习惯了被照亮的感觉，而为了逃生进入黑暗的墙角，竟然有点不适应。我慌张得找不到路，幸好遇到一位好心的同类。

After a while spent under the lamp, I became used to the light so it was an uncomfortable feeling to escape back into the darkness. I was so flustered in fact that I completely lost my way. Luckily, I bumped into a fellow spider.

他是一只幽灵蛛。和爱光的小飞虫相反，他们总喜欢待在黑暗且不被注意的地方。在黑暗中行走是幽灵蛛的特长，他在前面带路，终于帮我顺利地从虎口脱险。

He was a daddy longlegs. Unlike the flying insects, daddy longlegs prefer to stay in dark and hidden places. Walking in the dark is their specialty and this daddy longlegs led me through the gloom and out of danger.

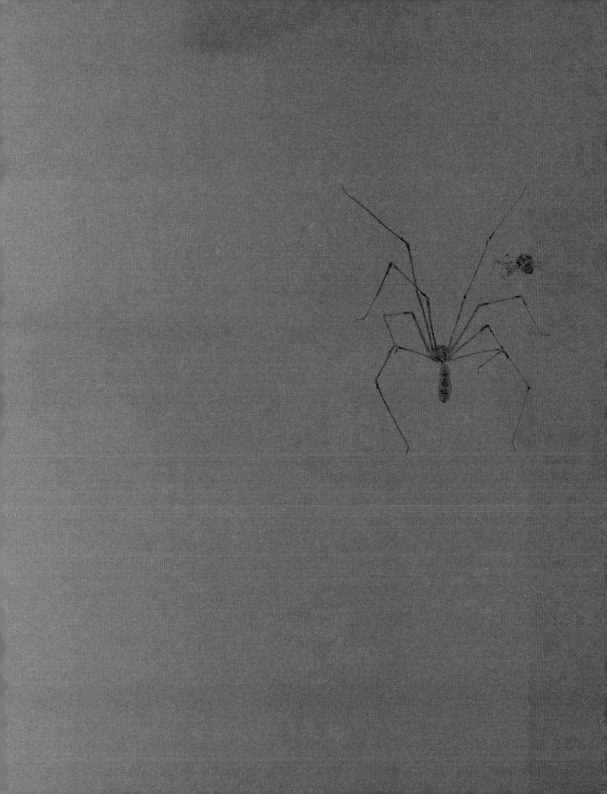

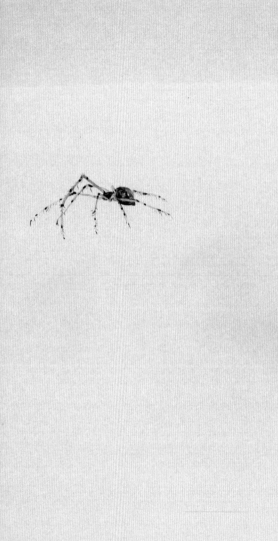

幽灵蛛改变了我对黑暗的看法。当我在黑暗中安静下来，黑暗便不再让我恐惧，反而给了我安全和保护，也给了我休息和放松的空间。

The daddy longlegs completely changed my perspective on darkness. When I finally felt calm in the darkness, I was no longer scared of it. In fact it even gave me a sense of safety and protection, a place to rest and relax.

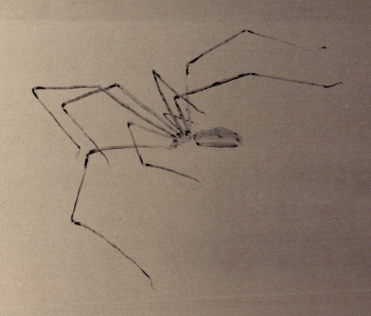

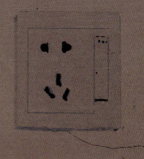

因为不同的习性，分离在所难免。离开了幽灵蛛之后，我第一次体会到了强烈的孤独，渴望有朋友做伴。某一天，一个红色的身影瞬间吸引了我。

Living different ways of lives, it was inevitable we would eventually part ways. After leaving the daddy longlegs, I felt incredibly lonely for the first time ever. I longed for the company of a friend until one day I suddenly caught sight of a bright red creature.

我第一次看到红色的蜘蛛，赞叹她有如此惊艳的外表。她告诉我，她真实的名字叫"叶螨"，只是外形像蜘蛛，但并不属于蜘蛛家族。面对我想亲近的渴望，她平静地回应道：你看到的只是我的外表，你在我身上投注的只是你头脑中的想象和幻影。

It was the first time I'd seen a red spider. I was taken aback by her incredible appearance. She told me that she was a leaf mite and although she looked like a spider, she didn't in fact belong to the spider family. I was anxious to get a better look, then she said to me calmly. 'What you see is only my outer appearance. You are gambling on an illusion, an imagination of your own making.'

虽然我有些失落，但我知道自己找错了方向。善意的叶螨介绍我认识了另一位蜘蛛朋友——梅氏新园蛛。她和叶螨一样拥有姣好的容貌，身体散发着绿宝石一样的光芒。她的美让我有些自惭形秽，她仿佛看透了我的心思，友好地对我说：每一个生命都有值得欣赏的地方，你也一样。

Although I was a little lost, I knew where I'd gone wrong. The kind hearted leaf mite introduced me to another spider friend—the spotted orb—weaver. Like the leaf mite, she was also incredibly striking and glimmered like an emerald, Her beauty intimidated me and then, as if looking right through me, she said sweetly: 'Every soul has its purpose, and so do you.'

梅氏新园蛛问我为何要找朋友，我说因为害怕孤独。她说：也许你需要一位老师。她带我结识了一只特殊的蜘蛛——大腹圆蛛，我早已久闻他的大名。

The spotted orb-weaver asked why I was looking for friends. I told her because I don't want to be alone. 'Perhaps you need a teacher,' she said, then took me to meet a very special spider, the big-bellied orb-weaving spider. It was not the first time I had heard of him.

· 68 · 69 · 大腹圆蛛是蜘蛛家族中的巨无霸，他不仅拥有雄傲于世的庞大身躯，更是织网的顶级高手，他可以把网织得又大又圆，规则又好看。不像我的网，总是毫无章法的混乱一片。我决定拜他为师，向他学习织网的技术和生命的道理。他慈爱地向我伸出手，邀请我进入他的家门。

The big-bellied orb-weaver is among the largest in the spider family. He had an impressive, rotund body and was an expert web weaver, making great big circular webs that are evenly balanced and beautiful, quite different to my rather more chaotic webs. I asked if he would teach me about web weaving and morals in life. He so kindly extended a hand and invited me into his home.

· 70 · 71 · 大腹圆蛛手把手地教我如何织网，指导我从第一根丝织起。虽然我有些笨拙，但他用温暖的鼓励不断地给我信心。

The big-bellied orb-weaver meticulously taught me how to weave a web, starting from the very first strand of silk. I may have been a little clumsy, but he fuelled my confidence with gentle encouragement.

在大腹圆蛛的悉心教导下，我终于学会了顶级的织网技术。看到自己会织出又大又圆的网，我们师徒二人兴奋又开心。他的大网总是能粘住很多虫子，有大也有小，而他只吃那些大的虫子，把所有的小虫都送给我作为学习的奖励。

Under his careful instruction, at last I mastered the most advanced web weaving techniques. Both of us were equally delighted that now I could weave big round webs. His webs were always full of different insects. He took the big ones for himself and gave the little ones to me as a reward for my progress.

· 74 · 和大腹圆蛛在一起的生活幸福又安全，我真想这样生活到永远。但大腹圆蛛告诉我，每个蜘蛛都有自己
的使命，不可以逃避。告别了大腹圆蛛，我下定决心从此依靠自己。

I was happy and safe with the big—bellied orb—weaver and felt I could live like this forever, but he told me that each
spider has its own mission that they must face up to. I said goodbye to the big—bellied orb—weaver, knowing that from
now on I must fend for myself.

· 75 · 我牢记大腹圆蛛的教诲,精确地测算距离,控制好节奏和疏密。织网时全身心地投入并保持专注,渐渐进入一种忘我的状态, 内心充满了幸福感。

I carefully memorised everything the big-bellied orb-weaver told me—how to measure distances with accuracy, how to control the rhythm and density—and as I concentrated intently on weaving, I slowly began to lose myself to the point of utter happiness.

· 76 · 77 · 我知道我只是普通的温室希蛛，我们通常只能织很小的网。但这一次我有足够的自信，终于像大腹圆蛛那样成功织出来一张大网。

I may be a just common house spider that makes tiny little webs, but with enough confidence at last I wove a great big web just like the ones made by the big-bellied orb-weaver.

·78·79· 天有不测风云，当我正陶醉在织出大网的成就感里，忽然来了一场暴雨，毁掉了我的大网。我虽心有不甘，但也庆幸腹中的储备还可以维持一段时间，我还可以东山再起。

But then something happened. There I was revelling in my web-weaving success when all of a sudden there came a great downpour that destroyed my web. It was all just too much. I was grateful at least to have reserves in my stomach that would keep me going for a while. I knew I would make a comeback.

　某一天，一个细长腿的身影出现在眼前，我以为是幽灵蛛，按捺不住重逢的喜悦。我正诧异他为何白天也出来闲逛，走近才发现，这是长相貌似幽灵蛛的盲蛛。盲蛛和叶螨一样，都是貌似但实非蜘蛛，他们不织网，像跳蛛一样靠捕猎为生。

One day, a creature with long slender legs appeared in front of me. At first, I thought it was the daddy longlegs from before and I couldn't contain my excitement to see him again. I wondered why he was out in the daytime, but as he grew closer I realised it was in fact a harvestman. Harvestmen, like leaf mites, look very similar to spiders even though they are not. They cannot weave webs and, like jumping spiders, they are hunting creatures.

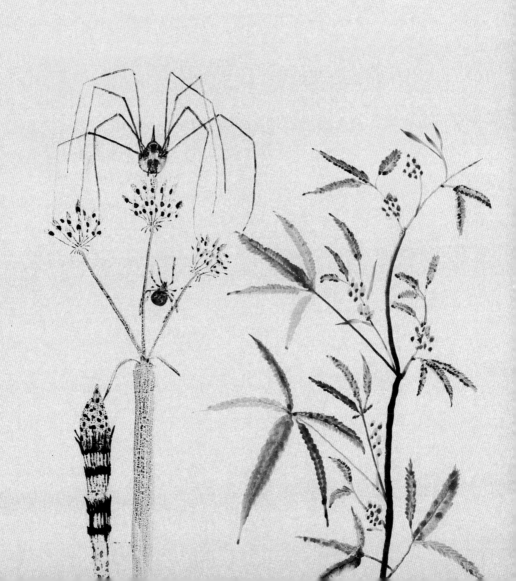

夜晚来临，一大群盲蛛集体出动去猎食。他们鼓动我加入他们的队伍，这一次，我没有动心。
我知道我和他们不同，我只想做我自己。

As night fell, a big group of harvestmen gathered together to go hunting. They encouraged me to join them,
but I wasn't interested this time. I knew that I was different from them. I just wanted to be myself.

历经风雨的磨砺，我的心已经变得越来越有韧性。我想起大腹圆蛛说，每个蜘蛛都有自己的使命。我打算织一张比之前更大的网，创造普通温室希蛛织网的奇迹。

Through trials and tribulations, my heart grew ever more resilient. I remembered what the big-bellied orb—weaver had said: each spider has its own mission. I decided to weave a web bigger than ever before, something miraculous for a common house spider like myself.

我一边不慌不忙地织网，一边想象着可能到来的收获。这时，远处飞来一只天牛，我知道这是一个危险的劲敌，但危险之中常常蕴藏机遇。

As I weaved with great care, I imagined all the things I could catch. Just at that moment, in flew a long-horned beetle. I knew this was a dangerous opponent, but also that danger so often brought with it opportunity.

天牛根本没把我这个小不点放在眼里，肆无忌惮地冲上来，对我的网拳打脚踢。

The long—horned beetle didn't pay any attention to me at all. It bounded over recklessly then proceeded to punch and kick my web.

· 89 · 我在蛛网上和天牛奋力周旋，好几次险些被天牛的利牙咬住，尖爪刺伤。我小心躲闪避让，天牛抓不住我
但又脱不了身，变得慌乱又躁狂。

The long-horned beetle and I had a standoff in my web. I narrowly avoided being bitten several times or being stabbed with
his sharp claws. I dodged and escaped with great agility and although the beetle couldn't catch me, he also couldn't get
away. The long-horned beetle became flustered.

我不知道哪来的勇气和灵感，瞄准天牛的一根触角，用蛛丝缠住。单根的蛛丝虽不起眼，拧成一股绳之后就能创造奇迹。天牛继续横冲直撞，越挣扎打转，就被蛛丝缠得越紧，终于耗尽了力气。

I don't know where the courage or inspiration came from, but I saw my moment and grasped one of the beetle's antennae, wrapping it in my silk. A single strand of silk might seem like nothing, but several strands wound into a rope could have miraculous power. The long-horned beetle continued to rampage, tossing about until it became so tightly entangled that it lost all strength.

·92·93· 捕获的天牛又给我带来了丰富的存粮。我让自己放松下来，有了闲适的心情，终于可以去散步看风景。阳光下，一颗毛茸茸的丝球闪闪发亮，白色的丝中间还包着一粒绿宝石。我很好奇，大胆地走上前，想看个究竟。

The long-horned beetle catch would keep me going for a while, so I allowed myself to relax. Feeling at ease, at last I could go out and enjoy the scenery. Beneath the sunlight I saw a hairy ball of silk glistening. Wrapped inside the white silk ball was an emerald. Curious, I ventured forward to get a better look.

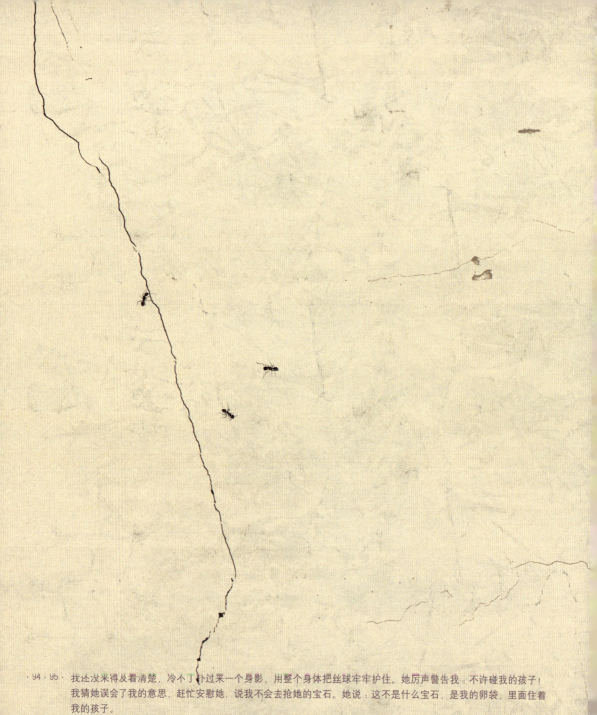

我还没来得及看清楚，冷不丁扑过来一个身影，用整个身体把丝球牢牢护住。她厉声警告我：不许碰我的孩子！我猜她误会了我的意思，赶忙安慰她，说我不会去抢她的宝石。她说：这不是什么宝石，是我的卵袋，里面住着我的孩子。

Before I had a chance to see clearly, another creature suddenly popped out, shielding the silk ball with its body. 'Don't touch my children!' she warned me sharply. I think she misunderstood me and I promptly assured her that I had no intention to steal her gems. 'These are not gems,' she said, 'this is an egg sack full of my children.'

我友善的态度让她放松下来，她开始和我交谈。原来她叫"肖蛸"，是蜘蛛家族的又一个特殊种类。他们有着比梅氏新园蛛更美丽的外表，人类因为贪恋他们的美貌，喜欢捕捉他们当作宠物饲养。肖蛸告诉我，近期周围常有寄生蜂出没，一定要多加小心。

Sensing my friendly attitude, she became relaxed and began to chat. She was a long—jawed orb—weaver, another special species in the spider family. These spiders are even more beautiful in appearance than spotted orb—weavers. People like to keep them as pets and admire their beauty. The long—jawed orb—weaver told me there had recently been an influx of parasitican wasps and that we must be careful.

我时刻记着肖蛸的提醒，防备周围的一切。有一位不速之客尾随而来，我开始盘查他的底细。他傻傻地看着我，语无伦次的样子有些滑稽。我忽然有了异样的感觉，听到了自己的心怦怦乱跳，竟然不知道手脚如何安放才好。我不但放下了对他的防备，还产生了想要接近的冲动，这种感受真的很奇妙。原来，这就是"爱情"。

With the long-jawed orb-weaver's warning in mind, I kept a look out in every direction. It appeared I had acquired a stalker and I tried to look behind me to suss things out. He looked at me in a funny way with a confused expression that was almost comical. I suddenly had a strange feeling, I could feel my heart pounding and didn't know where to place my hands and feet. Letting my guard down, I felt compelled to approach him. The feeling was wonderful. The feeling, it seems, was love.

我和他相依相伴在一起，一刻也不想分离。蜘蛛家族一直有种传统，为了保证雌蜘蛛有足够的营养繁衍后代，雌蜘蛛通常在交配成功之后把她们的雄性伴侣吃掉。我觉得这样的传统有些残忍，更觉得爱不该是占有和牺牲。我相信自己有足够的能量养育孩子，不想吃他，而是想和他共度余生。

Once we were together in one another's company, the idea of separating even for one second seemed unimaginable. In the spider family, there is a longstanding tradition whereby female spiders often eat their male partners after mating to give them the nutrition they need to reproduce. I think this tradition is a little cruel. Love should not be about domination and sacrifice. I had plenty of energy to raise my children and no desire to eat him. I wanted to spend the rest of my life with him.

我们终于有了爱情的结晶，我沉浸在即将为母的喜悦中，他却因为畏惧那个传统，弃我而去。我像所有的蜘蛛妈妈一样，为了抚育孩子，倾尽全力。我用蛛丝给他们织了一个牢固的卵袋，里面铺上细细的绒毛被，既能保暖又能挡风遮雨。我把所有的心思都倾注到孩子们身上，竟然忽略了危险已然逼近。

We were in love and not long after I was immersed in the joy of motherhood. But even still he abandoned me for fear of that tradition. Like all spider mothers, I put everything into raising my children. Using silk, I wove a sack for my eggs, and covered it with a thin fluffy quilt to keep them warm and protected from the wind and rain. I devoted myself to caring for my children, so much so that I was unaware of the danger that approached.

一只寄生蜂闻风来袭，我想起了肖蛸的提醒，但想躲藏已经来不及。为了编织卵袋，我几乎吐尽了腹中所有的蛛丝，耗尽了身体储存的能量。抱着沉重的卵袋，我也跑不快。眼看大难即将临头，我已无路可走，只得摆出迎战的姿态。

A parasitican wasp swooped in, reminding me of the long-jawed orb-weaver's warning, but it was too late to hide. Having used all the silk I could muster up to make my egg sack, I had no energy reserves left. Carrying the heavy egg sack, I couldn't run fast either. Disaster was ahead but I had nowhere to go. All I could do was stand there with an aggressive glare.

为了保全孩子们的性命，我想假装丢下卵袋逃跑，以此引开寄生蜂，又怕她伤害卵袋，左右为难，一时间没了主意。在我心里，孩子们的生命重于一切。我此刻深深地理解了肖蛸的行为，只要能保住孩子，就算是死也在所不惜。

I thought about pretending to abandon my egg sack to try and drive the wasp away and save my children, but I was afraid she might attack my eggs. I was in a tricky dilemma and for a moment I had no idea what to do. I knew in my heart that my children's lives were more important than anything. I empathised with the long-jawed orb-weaver's reaction from before. Even death seemed insignificant when it came to protecting my children.

既然死亡已不可避免，与其恐惧，莫如勇敢迎接。我使出全身的力量，用身体死死遮挡住卵袋，后背彻底暴露在外。寄生蜂毫不费力地用产卵器刺中了我的后背，把她的卵注入我的身体。这是寄生蜂的习性，等她的卵在我的身体里长大，那时我就会慢慢死去。

Death was inevitable, so I was better to meet it with bravery rather than with fear. Using my whole body, I shielded the egg sack, exposing my back completely. The parasitican wasp pierced its ovipositor into my back just like that, planting her eggs inside my body, as is the way of these wasps. Her eggs would grow bigger inside me and eventually kill me.

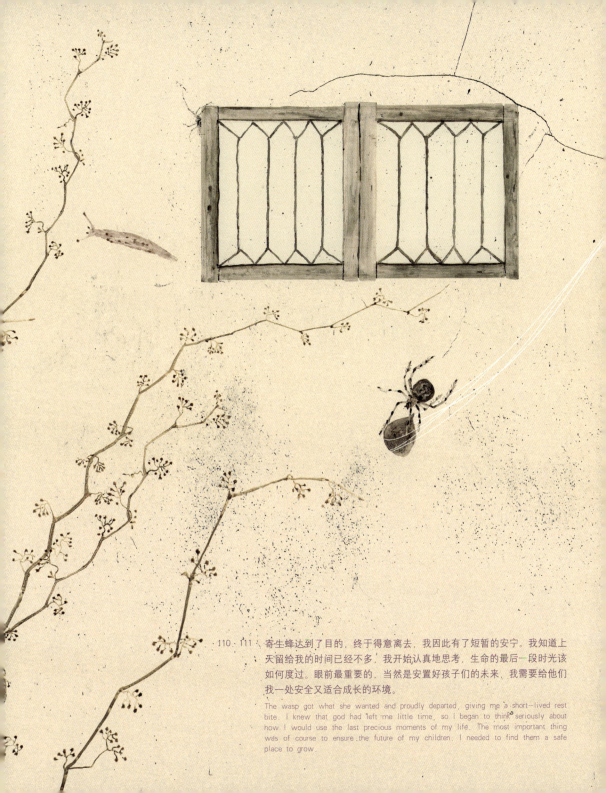

寄生蜂达到了目的，终于得意离去，我因此有了短暂的安宁。我知道上天留给我的时间已经不多，我开始认真地思考，生命的最后一段时光该如何度过。眼前最重要的，当然是安置好孩子们的未来，我需要给他们找一处安全又适合成长的环境。

The wasp got what she wanted and proudly departed, giving me a short-lived rest bite. I knew that god had left me little time, so I began to think seriously about how I would use the last precious moments of my life. The most important thing was of course to ensure the future of my children. I needed to find them a safe place to grow.

我找到一处避风的墙角，幸运地发现一片残存的蛛网，网上挂着一片大大的叶子，正好可以挡风避雨，便在这里给孩子们安好了家。我知道，剩下的时间就只能静静地等着孩子们自己长大，能不能见到他们，我甚至都无法确定。

I found a corner sheltered from the wind and luckily enough there was the remnants of a spider's web with a large leaf hanging on top. The leaf would make good shelter and so I settled my children down there. I knew that in my final days all I could do was wait for my children to grow, even if I wasn't sure whether or not I would meet them.

寄生蜂的卵在我的身体里渐渐长大，她的幼虫开始啃噬我的身体，我因为疼痛和虚弱一次次地晕了过去。半梦半醒之中，我竟然看到了自己的妈妈，她对我说：亲爱的孩子，你是一位了不起的蜘蛛母亲，你已经完成了你的使命，我会在天堂的家等你回来重聚。

The eggs of the parasitican wasp grew bigger inside me and the larvae began to chew away at my flesh. I felt faint under the pain and weakness and dipped in and out of consciousness. Half dreaming, half awake, I saw a vision of my mother. "My dear child," she said to me, "you are a remarkable spider mother and you have completed your mission in life. I'll be waiting for you in heaven."

我的身体一天天虚弱，保持清醒的时间越来越少。令我惊喜的是，伴随着寄生蜂的幼虫慢慢长大，我的孩子们也在渐渐长大。这一天，最顽皮的那个孩子率先咬破了卵袋，一群孩子纷纷钻了出来。我终于见到了他们，我最亲爱的孩子们。我竟然忘记了疼痛，眼里涌出的泪水，我已经无法分辨其中的含义。

My body grew weaker by the day and began to loose consciousness. To my surprise, as the parasitican wasp larvae grew bigger so did my own offspring. That day, the most cunning of my children bit a hole in the egg sack and out came a flood of children one after another. I finally met them, my dear children. For a moment my pain was gone, I was overwhelmed and tears welled up in my eyes.

孩子们围着我叽叽喳喳地说个不停，像所有的孩子一样，缠着我给他们讲故事听。我深深地吸了口气，用生命中最后一点力量，给他们讲了一个故事：从前，有一只小蜘蛛，她的名字叫"温室希蛛"，她长着小小的身躯，灰灰的面孔，看起来普通又平凡……

Gathering around me my children chirruped and chattered. Like all children, they pestered me to tell them stories. I took a deep breath and used the last bit of energy I had to tell them a story. Once upon a time there was a little spider, she was a common house spider with a small body and a grey face. She looked perfectly ordinary…

·120·121· "蜘蛛的一生仿佛注定要在两难之中徘徊。如果不吐丝织网，就无法猎食谋生；如果觅不到食物，腹中空空，就无法吐丝织网。如果网太小，就粘不到足够的猎物；如果网太大，又消耗太多的体能。网织得太高，容易被风刮跑；网织得太低，容易被水淹到。究竟怎样做才好，并没有唯一的真理。我知道你们不想离开，但只有学习去独立探索，才能找到属于你们自己的答案。"

It seems that spiders are doomed to a life of contradictions. If they do not make silk to weave webs, they cannot catch food to survive. If they cannot find food, they lack energy to make silk for their webs. If their webs are too small, they cannot catch enough prey. If their webs are too big, it takes too much energy. If their webs are too high, they are easily blown away in the wind, too low, and they are easily flooded. There is no single truth about what is best. I know you don't want to leave me, but only by learning and exploring on your own will you find answers.

我的故事讲完了，孩子们也听懂了我的嘱托，他们依依不舍地和我道别，开始了独立生活。回顾自己的一生，此刻我已了无遗憾。现在，我可以心无挂碍地吐出最后一口气，永远地闭上眼睛。

And here my story ends. Taking my advice, my children set out on their own journeys. Looking back on my life, I had no regrets. Now I could take my last breath unhindered and close my eyes forever.

我终于回归生命之源，化为尘泥。我的身躯虽然消失，但我已将自己融于整个天地。亲爱的孩子，我要告诉你：虽然我已不在，但我的爱会永远留下来！

In the end I returned to the dust and earth from which I came. Although my body no longer existed, I became a part of heaven and earth. Dear children, I want to tell you that although my body is gone, my love will remain here forever!

书中的故事情节在随园书坊现实生活中时有发生

温室希蛛在照顾卵袋和刚出卵袋的小蜘蛛

A common house spider taking care of her egg sack and watching her babies hatch

一只蜗牛正通过一根蛛丝向对岸滑行

A snail gliding across a spider's silk trying to get to the other side

墙壁上的梅氏新园蛛

A spotted orb—weaver on the wall

竹篱笆上，一只漏斗蛛埋伏在竹管里

A funnel—web spider lying in ambush inside a bamboo stalk on the fence

悦目金蛛正在织带有字母的网

A priest spider weaving a web filled with lettering

肖蛸正搬着白色卵袋在墙壁上艰难行进

A long-jawed orb-weaver carrying a white egg sack along the wall

敏捷的跳蛛在荷叶边缘，警觉地看着周围

An agile jumping spider sitting on the edge of a lotus leaf alert
and looking around

蛛蜂偷袭了一只漏斗蛛，并注射毒液将其麻醉

A spider wasp sneaking up on a funnel-web spider and
injecting it with poison

作者简介
The Information of Author

周宗伟 博士

南京师范大学心理学院副教授

主要从事艺术治疗、心理健康教育等相关领域的实践与教学研究工作，是国家二级
心理咨询师，美国 NGH 催眠协会注册催眠师。著有《高贵与卑贱的距离——学校文
化的社会学研究》。

Zongwei ZHOU, PhD

Associate professor, School of Psychology, Nanjing Normal University

Her recent research has focused on art therapy, mental health education and other
related fields. She is a National Psychological Consultant (Level 2) and NGH Certified
Hypnotherapist. And the author of *The Distance Between Noble and Humble: A
Sociological Study of School Culture*.

朱赢椿

南京师范大学书文化研究中心主任，全国新闻出版领军人才

由他创作与设计的图书数次被评为"中国最美的书"和"世界最美的书"。2008 年，
《蚁呓》被联合国教科文组织德国委员会评为年度"最美图书特别制作奖"。2017 年，
《虫子书》荣获年度"世界最美的书"银奖，并被大英图书馆永久收藏。

以《虫子书》为代表，《虫子旁》《虫子本》《虫子诗》相继出版——微观世界，蔚为大观。
同时，"虫子说"系列也逐渐走向世界，《蚁呓》《蜗牛慢吞吞》《蛛嘱》先后版权输
出至韩国、捷克等国。2022 年，《蜗牛慢吞吞》还被改编为舞台剧在德国上演。

理想国还出版有朱赢椿作品《设计诗》《语录杜尚》《便形鸟》。

Yingchun ZHU

Director of Book Culture Research Center, Nanjing Normal University

National Press and Publication Leading Talents

His books and book design works have won and been nominated 'Beauty of Books in
China' Award and 'Best Book Design from all over the World' (Stiftung Buchkunst) many
times. In 2008, the UNESCO German Commission identified *ANT* as that year's 'Most
Beautiful Book'. In 2017, *THE LANGUAGE OF BUGS* won the Silver Medal Winner in 'Best
Book Design from all over the World', and was in the permanent collection of the British
Library.

To be by bugs side, Yingchun ZHU created a BUG UNIVERSE by a series of bug-
oriented books.

Beijing Imaginist Time also published other works by Yingchun ZHU such as *POETRY IN
DESIGNED, QUOTATIONS FROM DUCHAMP, CACAFORM BIRDS*.

图书在版编目(CIP)数据

蛛嘱 / 周宗伟著；朱赢椿绘. -- 桂林：广西师范
大学出版社, 2022.6
ISBN 978-7-5598-4812-3

Ⅰ.①蛛… Ⅱ.①周…②朱… Ⅲ.①人生哲学 – 通
俗读物 Ⅳ.①B821-49

中国版本图书馆CIP数据核字(2022)第042795号

广西师范大学出版社出版发行

广西桂林市五里店路9号　邮政编码：541004
网址：www.bbtpress.com

出　版　人：黄轩庄
责任编辑：马少匀　董婧
装帧设计：朱赢椿　小羊
全国新华书店经销
发行热线：010-64284815
北京华联印刷有限公司印刷

开本：889mm×1194mm　1/24
印张：6　字数：5千字　图片：76幅
2022年6月第1版　2022年6月第1次印刷
定价：69.00元（精装）

如发现印装质量问题，影响阅读，请与出版社发行部门联系调换。